THE EDWARDIAN FARM

Jonathan Brown

SHIRE PUBLICATIONS

Published by Shire Publications Ltd,
PO Box 883, Oxford, OX1 9PL, UK
PO Box 3985, New York, NY 10185-3985, USA
Email: shire@shirebooks.co.uk   www.shirebooks.co.uk

First published 2009.
Transferred to digital print on demand 2015.

A CIP catalogue record for this book is available from the
British Library.

Shire Library no. 490   •   ISBN-13: 978 0 7478 0714 8

John Sandon has asserted his right under the Copyright,
Designs and Patents Act, 1988, to be identified as the
author of this book.

Designed by Ken Vail Graphic Design, Cambridge, UK
Typeset in Perpetua and Gill Sans.
Printed and bound in Great Britain.

FSC
www.fsc.org

MIX
Paper from
responsible sources
FSC® C013604

COVER IMAGE
A selection of Worcester porcelain. From left: figure of
'Joy', dated 1913; a cup by Thomas Baxter, c. 1815; a vase
by Walter Austin, 1911; a blue and white teapot, c. 1770;
a scale blue pattern vase, c. 1768; and a 'painted fruit' vase
by William Roberts, c. 1965.

TITLE PAGE IMAGE
A Barr, Flight and Barr vase painted with shells, attributed
to John Barker, 33 cm high, c. 1810.

CONTENTS PAGE IMAGE
A pair of Eastern Watercarriers modelled by James Hadley
and decorated by Edouard Béjot using metallic enamels and
old ivory. 24 cm and 25 cm high, factory marks dated 1881.

ACKNOWLEDGEMENTS
In the great majority of cases, where no individual
acknowledgement is given, illustrations show pieces sold
by Bonhams. This book would not have been possible
without access to the precious archive of pieces that have
passed through my hands during thirty-three years as a
specialist at 101 New Bond Street, London W1.

CONTENTS

INTRODUCTION

THE REIGN of King Edward VII was a short one, from 1901 to 1910, but it quickly gained the impression of being a special time, a golden age even, which lasted to the onset of the First World War in 1914. To a later generation it seemed a time of peace and prosperity. Many Edwardians themselves thought in terms of embarking on a new age coinciding with the new century. The character of the King himself, flamboyant in comparison with his mother, imparted something to this air. In many ways it was a prosperous time, for which the phrase 'never had it so good' might have been coined, and, viewed in retrospect from the 1920s, incomparably so.

For rural England, this seemed a perfect age of country house parties, the upkeep of old landed traditions, while sharing in some of the modern luxuries such as motoring. It was the country life that entered later literature, such as L. P. Hartley's *The Go-between*, published in 1953 but set in the summer of 1900.

The farmers shared in this to some extent. A. G. Street was a farmer well known as a journalist between the wars and later. Looking back from the 1930s to his youth on his father's 600-acre farm on the Wiltshire downland, he could describe these Edwardian years as 'the spacious days', all pleasure and sport. As tenants of a large farm, the Streets, father and son, were invited by their landlord to join the estate shooting parties. Mr Street was in a position to act more as manager than hands-on farmer. He was not alone in that, but the experience of one of the other farmers featured in these pages was more typical. Working a much smaller, mainly dairy farm, there was no sign of spare time in the average day; just a steady round of toil.

Opposite:
Stacking the hay.
Despite the
introduction of
steam and motor
power, most of the
work in this period
was still done by
men and horses.

LAWES'

MANURES

THE VICTORIAN LEGACY IN FARMING

EDWARDIAN FARMING was shaped by its inheritance from the Victorian era. This was a two-fold legacy. In the first place, agriculture had become more productive during the nineteenth century. Alongside the industrial growth in the British economy, farming had been through its own revolutions that had set the pattern still followed by Edwardian farmers. Intensive rotations of crops to get the most out of the land was one feature of this trend. Farmers were anxious to put goodness back into the soil, and increasing quantities of fertilisers were bought in to add to the manure produced on the farm. Crushed bones, guano, superphosphate (a chemical treatment of bones) and basic slag (a by-product of steel making) were among the new 'artificial' fertilisers used. Additional feedstuffs could be bought for livestock, mostly 'cake' made from linseed, rapeseed and similar products. The Edwardian farmer also had the first 'compound' feeds mixed by millers. The livestock treated to these feeds were of improved breeds – the Shorthorn cattle, and Leicester and Southdown sheep. Seed drills, steam engines and threshing machines had been introduced; even the 'basic' farm implements of ploughs and harrows were of improved engineering.

There was a second, more immediate part to the Edwardian farmer's inheritance. During the last quarter of the nineteenth century, farming had been through a period of severe economic recession – known as the Great Depression – and it had forced a lot of change on the farmers. Prices for agricultural produce had fallen heavily, and with them farm incomes. The farming industry as a whole had shrunk, with output down by about 13.5 per cent in England between 1870 and 1900.

Edwardian farming was in recovery from that depression. The corner of agricultural prosperity had been turned during the late 1890s, in that the prices of agricultural produce were no longer falling, and farmers' costs had been reduced, especially rent and labour. Farmers had a much better prospect of making a living from their soil; even so, as the new century opened, many farmers were far from optimistic – after a long period of trial who could blame them? Sir Henry Rider Haggard was well known as a novelist, but was

Opposite:
A romantic view of ploughing in an advertisement for Lawes' manures.

Victorian farming increased its use of imported and manufactured fertilisers, of which the company founded by Sir Joseph Lawes was a leading supplier. This is one of their leaflets from c. 1910.

also a landowner in Norfolk, and he conducted a tour of England's farming, which was published in 1902. Many of the farmers he visited were not inclined to look on the bright side, and nor was Haggard himself. His general conclusion about the state of English farming was that it was a 'failing industry'. He wrote, 'Many circumstances combine to threaten it with ruin, although as yet it is not actually ruined.'

Nevertheless, farming was on the road to recovery. Even Rider Haggard had to concede that there were farmers in many parts of the country, Shropshire and the Fens of Lincolnshire among them, who were farming well and profitably. Even in his own county of Norfolk there were farmers doing well: Lord Leicester told Haggard about a number of his tenants expanding their farms and getting good returns. A. G. Street wrote admiringly of his father's ability to make money from his Wiltshire farm — looking back from the 1930s when it was more difficult to make profits in downland. Although it was not plain sailing, throughout the Edwardian years production

One of many new implements of the Victorian era, the horse rake for gathering the hay crop had been introduced in the 1880s. This is Ransomes' rake of the 1900s, from one of their trade cards.

RANSOMES' HORSE RAKES

and profits on the farm were improving. One indication is the fact that farmers were prepared to invest in the more expensive technologies. Sales of steam cultivation equipment, mostly to the contractors hired by farmers, picked up during this period, while there was real interest in the development of the tractor. Sir Daniel Hall, another leading agriculturalist, toured a number of English farms, resulting in the publication in 1914 of

Oilcake, one of the mainstays of Victorian farming progress, was still important as animal feed in the Edwardian period.

A farmhouse at West Camel, Somerset c. 1912.

A brick farmhouse at Corringham, south Essex, photographed c. 1910. It was built in the early seventeenth century. There is a labourer's cottage attached on the right, the farm buildings on the opposite side.

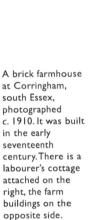

A Pilgrimage of English Farming 1910–1912. He concluded that farming 'is at present sound and prosperous'. Total agricultural production had recovered from its low point in the mid-1890s. It could be said to have found a new equilibrium, to have found its feet again after the years of poor results during the late-nineteenth century. But farmers, and some of the ways they farmed, were different now.

Often it was a new generation that had come into the farm, as the older ones had retired or given up in the face of a long period of falling prices for their wheat, wool and meat. This new generation was having to take a fresh look at how they farmed, to be more business-like, or at least business-like in a different manner. Some had come into farming with no experience, like the businessman who introduced the

Spraying potatoes, Cippenham Court Farm, Buckinghamshire.

King Edward potato to the nation. This was John Butler, who had been a grocer and draper at Scotter in north Lincolnshire, and went into farming in the mid-1890s. He bought the stocks of a new and hitherto unsuccessful variety of potato, which he renamed 'King Edward', a branding which turned failure into runaway success. By 1904, Butler had nine farms to his name. There were many like him. Indeed, the farmer who had entered the business from trade, 'who spent little, but made farming pay', as Rider Haggard was told in Dorset, became part of a new folklore.

The advent of motor transport and the telephone was starting to have its effects on the marketing of farm produce. Many livestock markets, such as this one at Petersfield, Hampshire, were still busy, however.

The Square, Petersfield

WHO WORKED THE FARMS?

W HEN EDWARD came to the throne in 1901 the number of farmers in England and Wales was about 225,000 according to the census of that year, a small fraction of the total population of 35.5 million. There were 620,000 agricultural labourers, and, taking into account the members of the farmer's family, the contractors and others engaged in agriculture, a total of about 1.2 million is reached. Farming was now a small part of a larger industrial, commercial and urban economy: no more than 8 per cent of the labour force of Great Britain, having fallen from 15 per cent in 1871. Numbers were falling as well: in 1871 there had been 962,000 labourers and 249,000 farmers.

Those 225,000 farmers embraced a wide range of people. Probably the majority had been born into farming and followed in family footsteps. Others came into farming from outside. Some of these were drawn from the ranks of farm labourers, but most entered farming from another line of business.

Some farmers were highly successful businessmen. Mr Street, with his 600 acres in Wiltshire, was one of them, but there were some who farmed on a scale larger than that. George Baylis, who farmed a total of 10,000 acres on the Berkshire Downs, was at the top of the scale. He had built up his enterprise, taking advantage of the opportunities presented by farmers who had fared badly during the depression, leaving farms available for him to take over, one after another. Others had large farming estates of more than 1,000 acres, such as William Dennis, who had 4,000 acres in the Lincolnshire Fens. Any farm of more than 300 acres was regarded as large at this time: the official statistics do not record any subdivisions beyond this. In 1908 there were 15,000 farm holdings recorded as greater than 300 acres, only 4½ per cent of the total number of 340,000. The equivalent in 2006 was 14 per cent of the total. Those who had the large farm holdings were the elite of farming society, and often wealthy with it: although in a small minority, they had a disproportionate influence, for their farms occupied nearly 30 per cent of the agricultural land of England, and in the eastern counties nearer 80 per cent.

Opposite:
A Devon farmer with an assured air, photographed at the beginning of the century.

A. Stratton had a large farm at Alton Priors, Wiltshire. Farmer and workmen pose behind a flock of sheep, c. 1910.

Most Edwardian farmers operated on a small or medium scale. Of the 340,000 farm holdings of more than 5 acres in England and Wales in 1908, more than half (58 per cent) were smaller than 50 acres; 37 per cent were between 50 and 300 acres. The very small farms under 50 acres were in a special category in many ways. Many were part-time holdings, but there were small dairy farms and small horticultural farms of the Fens or Vale of Evesham that could provide a living for a family. The 'typical' farmer, if one may be found, was the farmer of 100–200 acres or so.

What nearly all Edwardian farmers shared, whatever their background or scale of business, was that they were tenant farmers, as their fathers and grandfathers had been before them. In broad terms about 90 per cent of the farms were tenant farms. Farmers rented their farms from landowners: some of them aristocratic and greater gentry owners of many thousands of acres; others the local squire with smaller estates; and some corporate owners, such as the Crown Estate or Oxford colleges.

Many of the farms recorded as owner-occupied were in the landowners' hands because they could not get a tenant. A few landowners had made a positive decision to farm their own land. Lord Wantage took this step in the 1890s, taking 10,000 acres into direct management at Ardington and Lockinge in Berkshire. Lord Rayleigh ran some of his land as an estate enterprise managed by his brother, Edward Strutt. These were rare examples. Of greater effect was the sale of surplus land that some estates were embarking upon, and farmers were among the purchasers. Quite a few of the farms George Baylis took over were freehold. The result of these developments was that the balance was starting to change, so that by 1914 there was more owner-occupier farming. This was a movement that gathered pace after the First World War.

The Edwardian farm was predominantly a family farm. Most farmers were married, theirs being something of a partnership in the land. Single farmers were mostly widows or widowers. Even those with large farms will have thought of themselves as family farmers, although they were in a position to act as managers of a large number of employees. On A. G. Street's father's farm there was a 'crowd of men about the place'. There were seven carters in charge of the horses, six milkers for the dairy cows, two shepherds, six day labourers, a foreman and a groom – twenty-three people in all. Those were the regular employees; at hay-time and harvest there was additional casual labour. There was a distinct hierarchy: the horsemen and shepherds were at the top of the tree. They were usually employed by the year, on a contract starting at May Day or Michaelmas, depending on which part of the country they were in. Some were still taken on at the annual hiring fairs, but this was a practice in decline. Below these men were the general labourers – the day labourers, so called because they were employed on terms that offered them only the day's work at a time.

It had been common for the young single men in the jobs hired by the year to live in the farmhouse. This was less usual in the Edwardian period, but by no means unheard of. However found, accommodation was often provided at no cost or subsidised; there were other allowances in kind for fuel

A scene on Lord Wantage's farms on the open downland at Lockinge in 1904, with sheep feeding off a crop of clover.

and food which added to the basic cash wages. Even allowing for all that, wages were not high — less than a pound a week for the best paid, the shepherds, when a clerk in town could earn nearly half as much again. The general labourers were paid on average about 16s 6d a week, including allowances, to the 19s 7d of the shepherd. Low though these figures were,

Above: Everybody, including the foreman, lined up to be photographed in the harvest field when this crop of wheat was being cut. The machine is a reaper, which cuts the corn only, leaving the swath to be gathered by the man on the left for others to tie into sheaves and build into stooks. This was issued as a postcard in 1910.

Right: The carter with his horse team, Wincanton, Somerset.

Opposite: Portrait of a shepherd.

the Edwardian farm labourer – this was still the general term, although the name of the National Union of Agricultural Workers founded in 1908 indicated a new word was creeping in – was better off than his forebears. He was the beneficiary of the fact that prices had fallen since the 1870s by more than his wages, enabling him to afford more and better food. The quality of the cottages had also risen during the previous half-century, although there were still many who had to live in housing of a low standard.

Smaller farms needed fewer employees. For a sizeable minority, the only people in regular work on the farm were the farmer's wife and children, additional help being drawn in as required from neighbours and the pool of casual workers. The general report to the 1851 census had noted that a third of all farmers in the country employed no labour. Fifty years later nothing had changed in this respect: about 41 per cent of the farm workforce in 1908 were relatives of the farmers. Dependence on the family was most common in the pastoral west of the country, where many a small dairy farm could work without employing additional labour, but even in some parts of the eastern counties many of the smaller farms employed no regular labourers.

There were a number of independent agricultural workers who went from farm to farm doing such jobs as hedging, for which special skills might be required. Among these skilled itinerants were the sheep shearers. By the

Sheep-shearing gang in Wiltshire.

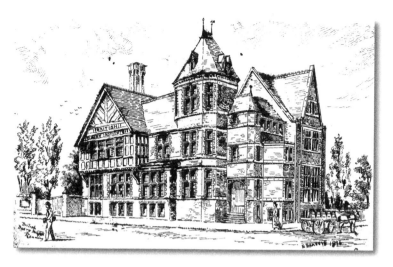

The new face of agricultural education: the British Dairy Institute at Reading, a drawing by E. Beattie in 1896, when the Institute was founded.

Edwardian years many of these were from New Zealand and Australia, who came to Britain to work during the antipodean winter. Rider Haggard described the arrival at his Norfolk farm of the shearers: 'four in number, who travel with a pony and cart from farm to farm, clipping the sheep at a charge that averages about threepence a fleece'. They were effectively contractors, although the term was used at this time for the men who owned the steam-powered machinery that was brought to the farm. There were threshing contractors and steam ploughing contractors; a few firms combined both roles. The contractors employed those directly involved with engine and machine: two or three men on a threshing set, and four or five on a ploughing set. The Edwardian period was something of an Indian summer for these contractors. Recovery since the 1890s had brought enough business for contractors to invest in some new engines and machines.

The Edwardian farmer was in general better educated than his predecessors. The young man entering farming would have passed through the national elementary education established during the 1870s. He might also have attended one of the new agricultural education classes and colleges. The Royal Agricultural College at Cirencester, founded in 1845, was still the most prestigious establishment, drawing its students mainly from the larger farmers and landowners. A movement to encourage technical education at the end of the nineteenth century led to the foundation of a number of new colleges for agriculture, such as Wye College in Kent (established 1894), the British Dairy Institute (1896) and Harper-Adams College, Shropshire (1901). There was a new text book for a new age: William Fream's *Elements of Agriculture* had first been published in 1892. It sold well and reached its seventh edition in 1905, the last one to be written by Fream himself.

WHAT THE FARMER GREW: THE MIXED FARM

MOST EDWARDIAN FARMS were mixed farms; that is, they had a proportion of arable land on which they grew some crops, and pasture land for livestock. The crops most of them grew were cereals, root crops and hay. Wheat and barley were the principal cereals, grown mostly as cash crops. The roots and hay were to feed livestock. The livestock they kept were principally cattle and sheep. In this nothing had changed from the farming of mid-Victorian times; indeed, mixed farming was to remain dominant until the middle of the twentieth century.

Most farmers thought mixed farming was best. It involved a mutual dependence of animals and crops, it was good for the land, and it gave the farmer more strings to his bow. Specialist farms were few. There were hill farms and lowland grazing farms that were almost entirely pasture, and there were a few specialist arable farmers. George Baylis grew little else but corn in almost continuous succession on his enormous farming enterprise in Berkshire. On the Fens there were some farmers who could be reckoned as specialist potato growers. But almost every form of husbandry was practised within predominantly mixed farming.

Everybody knew what mixed farming was, but defining it precisely was more difficult because what was in the mix varied greatly between different regions of the country and according to different types of soil. In 1851 Sir James Caird had famously drawn attention to the basic division between the mainly pastoral and livestock-keeping north and west, and the mainly arable and corn-growing south and east of England, and that still held good fifty years later. In eastern England, the great corn-growing region, the proportion of arable was likely to be three-quarters or more, and about half of that might be growing cereals, even after the long period of low prices for wheat. In the western half of the country, the proportions of land were reversed, and pasture was far more important. In Essex with its low rainfall, the clay soils were ploughed more than they were further west, in the clay vale of Wiltshire, for example. Cattle were an important part of farming in both places. The Streets farmed on the chalk in Wiltshire where they kept lots

Opposite:
The binder at work in the harvest field in Somerset. Though of a later date, this photograph shows an essentially Edwardian scene.

Right: Ploughing and rolling a large field at Eaton Green, Luton; a postcard of 1909.

Below: Sheep and lambs feeding in a close by the farmyard; the encapsulation of mixed farming. The large haystack in the process of being cut away for feed indicates a late-winter scene. A corn stack awaits threshing in the right distance.

The location of this farmyard is not known, and the photograph might not have won a prize in a competition, but it does give the flavour of a mixed farm, with the hens roaming, the enclosures for horses and cattle, the haystacks and the large modern barn.

of sheep and grew cereals: this was 'sheep-corn' husbandry, common on light-soiled uplands. All of these, and there were more regional and local types, counted as farming.

Mixed farming changed over time as well as from region to region. According to a writer in the *Standard Cyclopaedia of Modern Agriculture*, published about 1905, 'ordinary mixed farming is less common than formerly'. What he meant was that the mix in mixed farming had changed

Young and old at work harvesting potatoes.

Cattle in the farmyard at Wincanton, Somerset, c. 1900.

considerably since the 1870s. The recession of the late nineteenth century had forced a major rethink. Farmers were sensitive to the market, and its signals were clear. The price of wheat had fallen dramatically, by nearly 50 per cent; the other cereals, barley and oats, had also declined in price by about one-third. The products of livestock farming had fallen in price by smaller percentages: meat by 15–20 per cent, butter and cheese by about 15 per cent, milk hardly at all. The result was that, whereas the mid-Victorian farmer had counted wheat and barley as his principal products, the Edwardian farmer relied more on livestock.

The 1908 census of production revealed that two-thirds of farming's output was of livestock and livestock products. It was the increase in livestock products, more than outweighing a decline in arable crops, that accounted for the recovery in total agricultural output between 1896 and 1911. The importance of mixed farming perhaps disguised that. The ploughmen continued to work the land, the harvest retained its significance for the Edwardian farmer. Harvest supper in September was still a big celebration of the end of another farming year, at least on large farms. Yet there were some significant differences. Much of the harvest now went to feed the livestock: even some of the wheat and barley sold was bought by other farmers to feed their stock. This was quite a role reversal, for many a mid-Victorian farmer had thought of the livestock as primarily for the benefit of the arable crops, in the manure that they could contribute.

Cereals, and wheat in particular, had become so unprofitable that farmers grew much less of them. The acreage of wheat grown in England and Wales had been cut by more than half between the 1870s and the 1890s. There had been some recovery since, but, except in the best corn-growing districts of eastern England, wheat did not have the dominant place it once did. The acreage of barley had also declined, by 24 per cent by 1900, but oats had increased because of their value as livestock food. There was still a market for grain crops: biscuit bakers bought British wheat, the brewers liked British barley, but times had changed.

The decline in the acreage of cereals since the 1870s had been matched by an increase in pasture. Four million acres had been added to the total of permanent pasture in England and Wales by 1900, and that acreage continued on its upward path throughout the first decade of the twentieth century, from 12,203,000 acres in 1900 to 15,972,000 acres in 1910. To some extent that had sharpened contrasts between farming regions, for the farmer of the north-west found it easier to reduce his cereal crop, retaining a modest extent of arable to provide feed for his livestock, than did the farmers of the south and east. As well as the permanent pasture, there was temporary pasture, grass sown as part of the arable rotation and mown for hay. This also was more important to the farmer of the early twentieth century, increasing from a little over 6 million acres in England and Wales in 1900 to 6½ million in 1910.

Southdown sheep, one of the main breeds for sheep-corn husbandry, at Babraham Hall, Cambridgeshire in 1903.

The increased importance of grass led to attention being paid to the quality of seed mixtures. One of the standard texts on pasture management was written by Martin Sutton, of the seed firm. These are three types of grass suitable for permanent pasture: meadow fescue, Sutton's perennial rye grass, and red clover.

THE ARABLE ROUND

There had been many changes, but the cultivation of the land remained one of the primary occupations for all mixed farmers. The principle underlying their farming was the maintenance of a balance of crops and stock that would keep the land 'in good heart'. Most farm tenancy agreements contained a clause that basically said that the tenant must return to the land what he took out. That traditionally had meant that the farmer was not to sell off such things as straw, hay and manure, but was to use them to maintain the fertility of the land. By the beginning of the twentieth century landed estates were taking a more relaxed interpretation of that clause, but they certainly did not strike it out. Landlords were more likely to give permission for straw and hay to be sold, provided that material of equivalent fertilising value was brought on to the farm. That could be managed by buying artificial fertilisers, such as superphosphate, potash, nitrates and basic slag, the use of all of which had increased since the mid-nineteenth century.

Farmers did not need tenancy agreements to tell them to keep the land healthy. It was ingrained into their thinking. 'One didn't farm for cash profits, but did one's duty by the land,' observed A. G. Street of his father's generation.

In order to achieve that healthy balance on the land, the farmer grew crops in a rotation of cereals, grass and clover, and roots. The grass and clover (what farmers referred to as 'seeds') and roots provided pasture and feed for the livestock, which in turn manured the fields to boost the return from the cereals. It was a neat system, a virtuous circle, which had been devised during the course of the agricultural revolution of the late-eighteenth and the nineteenth centuries, and the farmer of the Edwardian years was still following these principles. There was more to it, of course: in order to add fertility to the soil, the farmer bought oilcake to fatten the livestock and boost the value of the manure, and he bought additional fertilisers, artificial and organic. Many thought that the pinnacle was the Norfolk or 'four-course' system: it was the 'most typical rotation', said Fream. It compressed the rotation into four years, one year each for winter-sown corn (wheat or oats), followed by roots, spring corn (barley and oats) and seeds (clover or grass)

C. Greenham was the ploughman at work at the top of Steart Hill, Somerset, c. 1912.

in the fourth year. Its origins were in farming on the light-soiled chalk and limestone regions of the Downs and Wolds, and it had the aim of getting the maximum from wheat, the principal cash crop. Strict adherence to it enabled about half the arable land to grow cereals each year, about a quarter of it under wheat. The basic principles held good across the whole of southern and eastern England where mixed farming was practised.

A. G. Street described the operation of the four-course rotation on his father's large farm in Wiltshire as 'unalterable as the law of the Medes and Persians'. However, few Edwardian farmers followed crop rotations as compact as that. Most soils could not take it, and the stresses of the recent past had forced changes to save cost, reduce dependence on wheat at its low market price, and to increase the number of cattle that could be kept. Most rotations were of five or six years, and some were extended up to eight. Even on the chalk soils rotations longer than four years were often preferred. The simplest difference was in the length of time given to the seeds. Instead of the one year for clover in the four-course, the temporary pasture was left for two, three or more years. But there were other permutations of crops depending on soil and markets.

Many of these rotations had been allowed to lengthen during the late nineteenth century. It was part of that restructuring in the face of falling prices that reduced dependence on wheat, instead having a longer period for the grass sown in rotation.

Breaking up stubble with a cultivator. Martin's Cultivator Co. of Stamford issued a postcard booklet promoting their products. This example was posted in 1909.

MARTIN'S PATENT CULTIVATOR BREAKING UP STUBBLES

Michaelmas in the autumn was when many farm tenancies began (11 October, or 'Old Michaelmas', was a common date). It was the season that marked the beginning of the agricultural year, certainly as far as the arable cycle was concerned. Autumn ploughing was the big job of the season, when the stubble from the cereal harvest was broken up and fields prepared for sowing winter corn. This was the peak time in demand for horse power: every plough team was out, two horses to each plough, three to plough heavy land. At an acre a day, the usual amount a single-furrow horse plough would accomplish, ploughing was a steady process that would occupy most of the season. For the farmer with 60 acres of arable, that represented the number of days, more or less, for autumn ploughing. Many fields were ploughed more than once, while the heavy cultivator was also used to break up some stubbles.

Joseph Stevenson was the farmer at Oldfield Farm at Baulking, in the Vale of the White Horse in north Berkshire, as it was then (it is Oxfordshire now). His income was derived mainly from selling milk, but his was dairying in a mixed farming context. He kept a diary detailing the day-to-day operations of the arable side of the farm. His record for 1909 shows how this form of mixed farming was managed. As soon as the wheat had been harvested, the work of spreading dung on the stubbles began in readiness for ploughing them. The plough moved on from the wheat fields to those that had cropped beans and peas. By the time he had reached the oat and barley stubbles for ploughing in November, he had already been at work with the

Drilling and harrowing winter wheat, Somerset.

Winter work. Cutting and laying a hedge, with a billhook as the main tool, 1908.

Another important winter job was maintaining drainage channels, and this was done by hand. This man is at work ditching in 1910.

cultivator and harrow ready for sowing wheat in the middle of that month.

Meanwhile, Stevenson had started on clearing ditches and trimming hedges, work which continued through the next few months, reaching a peak in January. During this time, too, the first of the roots were being cut. The root crop was important to farmers with all but the smallest head of livestock. Turnips, mangolds (mangels) and swedes were the mainstay of feeding. Their disadvantage was that they were expensive to grow. There were few labour-saving implements. Thinning out the seedlings and weeding the growing crop had mainly to be done by hand. So did much of the harvesting – pulling the plants up by hand, and chopping of the top and tap root, in the cold and damp of autumn and winter. Cost-cutting farmers since the depression years had been taking a harder line with them. Many reduced the quantity of roots they grew, and for what they did grow they sought higher-

yielding varieties. The total quantity of roots grown in England and Wales in the Edwardian years was about a third less than it had been in the 1870s. There were cheaper alternatives to feed the stock, ranging from home-grown grain and other crops such as beans and maize to buying feeds. The Edwardian farmer had a greater range of animal feeds available to him, from the long-established oilcakes to Bibby's and Spillers' prepared feeds.

Farmers had cut down on roots, but few had dispensed with them altogether: they were still, in the words of one Edwardian farming text book, of much 'cultural and feeding importance, and as a rule a comparatively certain crop'. In places where livestock was especially important, in Devon for example, roots could still make up more than a quarter of the arable acreage.

Manure being carted to the field, 1905. It was dropped off in little heaps to be spread out later.

Seedsmen were keen to promote their high-yielding varieties of roots for animal feed. Carters' catalogues included this graphic depiction of their swedes.

31

The manure suppliers were just as keen to promote their wares. This trade card advertisement issued by G. Hadfield & Co. Ltd. gives perhaps an exaggerated view of successful mangold growing...

The record of how much land Stevenson devoted to his root crops has not survived, but he certainly grew substantial amounts of all three types. For additional fodder crops he grew peas and beans, and some maize. Primrose McConnell in Essex also grew maize for his dairy cows, but this was still a relatively unusual fodder crop in the Edwardian years.

... but mangold clamps could be impressive without the advertising.

Joseph Stevenson was not a large grower of cereals, and he had finished his autumn ploughing by the beginning of December. Others with more to

A gang of women weeding roots, Akeld Farm, Northumberland.

plough took longer, but the tradition was to finish by Christmas. Of course, not everybody did: Primrose McConnell still had some ploughing to do in January. There was a practical point to the tradition, for the first weeks of the new year were likely to be colder, making the land too hard to work. These weeks were quieter, certainly for the horsemen, but not without activity. There were loads of muck to be taken out to spread on the fields, feed to be carted to the sheep folded outside, clamps of potatoes and roots to be attended, and threshing.

The first of the winter threshing at Oldfield Farm was at the end of November, when the barley was done. Corn was threshed by the machine brought along by the contractor, so the diary notes its arrival and the number of days worked. This was the first of several visits, most of them during winter.

Ploughing for turnips, Akeld Farm, Northumberland.

Cippenham Court Farm, near Burnham in Buckinghamshire, was farmed by the Headington brothers in the Edwardian years. Horse-hoeing is under way in about 1909.

Peas were threshed in early December; wheat, oats and beans at the end of January to the beginning of February. One or two ricks were left until later in spring. Threshing involved a lot of people. The contractor's men were few, only the engine driver, a second man and a mate. The farmer had to find all the hands to feed the machine, bag up the corn, and for the carting. In all there could be a dozen involved.

The winter work also included chaff-cutting – chopping straw up for horse-feed. This might be done by the threshing contractor, with his large steam-powered chaff-cutter, and, although Stevenson's diary does not say so, it seems probable that this was how the wheat straw was cut up at Baulking. Other chaff cutting was done with the farmer's own machine, powered by horse gear, or, if a small one, by hand.

As mentioned above, the early weeks of the year were not a time for ploughing, but during January dung was being carted to the fields. That was followed by the busy time of spring cultivations: the work of cultivating, rolling and harrowing in preparation for the spring-sown crops of cereals and clover. Sowing of beans and peas was done at Oldfield Farm at the beginning of April, followed by mangolds at the end of the month. Stevenson did not get the oats in until early May – in general, late March and early April were the time for sowing spring cereals. Some of these times were regarded as critical, especially for barley grown for malting. The cereals were all sown by seed drill, and this could be heavy work. The big drills that could manage twelve rows at a time needed three horses to pull them, although just the one man was enough to control the early-twentieth-century machines. Grass and clover were more likely to be sown broadcast. On some farms this was still done by hand, but there were two alternatives. One was the seed fiddle, a hand-operated mechanical seed distributor, and the other was the broadcast (or shandy) barrow. The barrow could be purchased either as a hand-propelled or horse-drawn tool.

No. 1002.   February, 1907.

BARROW SEED DRILLS.

MANUFACTURED BY

R. HUNT & CO., LTD.

May and early June in the fields at Oldfield Farm were a constant round of weeding of cereals and root crops, mainly by horse hoe, although the diary implies that some was done by hand. This was where much of the cost of growing roots was incurred. Singling and weeding was a continuous process throughout the season requiring many days' labour. Getting the thistles out of the meadow was probably hand work, taking several days. The early growth of cereals was rolled to give them extra strength. There was also the summer ploughing, at Oldfield of fallows ready for crops of turnips and swedes sown in early June. The first sowing of roots meanwhile was being singled. On many farms this was also the time to give the grass a top dressing.

At the end of May 1909, Joseph Stevenson went to Wantage to get a new mower. That was the first step in preparation for getting the hay in. A few days later in June, the diary records that the swath-turner was mended and the wagons greased. A few days after that, mowing the hay harvest began, continuing into the first ten days of July. Haymaking was as important to the

The broadcast seed drill, or shandy barrow, for sowing grass seed was most common in this hand version, although bigger designs for horse power were also made.

Haymaking In Hampshire.

Mowing the hay, Hampshire, from a postcard of 1911.

35

An Albion mower, from a catalogue cover of 1912.

Edwardian farmer as it had been for his forebears. Along with the roots, hay was still one of the most important foods for the livestock of the farm. Haymaking had been largely mechanised since the mid-nineteenth century. Horse-drawn mowers cut more than an acre an hour. The crop was turned

Raking hay, Wiltshire 1909.

by the horse-drawn haymaker, or tedder, and was gathered by a horse drawn rake. There were powered elevators and stackers as well, these driven by a horse gear.

Carting, stacking and thatching the hay ricks continued through to early August, while in other fields the turnips were singled, and ploughing of summer fallows continued. Later in August a day was spent 'getting the

The Paragon hay fork, made by G. C. Ogle, was introduced during this period. This crane for building the stack was usually powered by horse gear and was most popular in the northern counties – this scene is from Northumberland.

Stacking the hay, Mill House Farm, Horley, Surrey, c. 1910.

Stacking the corn at Ripple Farm, Godmersham, Kent, 1913.

binder ready', another indication that he was up to date with his machinery, using the binder not the older reaper. The machine was put to immediate use in the wheat harvest. Being mainly a livestock farmer, Joseph Stevenson's corn harvest did not represent the peak of activity that it did on farms growing cereals for market. His wheat was cut and carted in a couple of weeks rather than the month or so it took on many another farm. Harvesting the pea crop, grown to feed the dairy cows, followed quickly on the wheat, while the rest of August was spent in further cultivation of fallows and weeding turnips.

LIVESTOCK IN MIXED FARMING

Cattle and sheep were the main types of livestock kept in mixed farming, and they were kept for three main purposes. Some farmers bred stock for onward sale to a second group of farmers who fattened them for the butcher, and a third group kept dairy cows. Many farmers combined two, sometimes all three of those roles. The Street farm had sheep and dairy cows. The distinction between the breeder and the grazier who fattened the stock was not as sharp as it had been; this was because demand for meat from younger stock had led many a farmer to fatten the animals he bred.

Whatever path the farmer followed, livestock had become more important as a source of revenue. Where farmers had kept cattle fed on large quantities of cattle cake in order to produce dung for the arable fields,

Shepherds in the southern counties would camp out in a hut on wheels during lambing time. This photograph of a shepherd's hut in use was taken for *Country Life* at Babraham Hall, Cambridgeshire in 1903.

they were now paying attention to their yield of beef or milk, and the feeding regime was altered accordingly. Sheep were kept for their meat more than the wool; the price of wool had dropped by about half during the late-nineteenth century, and the market was still weak in the Edwardian years.

Dipping sheep, c. 1909.

Washing sheep before they were sheared was common practice, but not universal, as not everyone was sure that clean wool affected the price. On this Norfolk farm a pond is being used for the job in 1914.

In consequence, such breeds as the Lincolnshire Longwool, prized for their wool, lost popularity compared with the Southdown and Hampshire Down, which were good for mutton.

According to one labourer on A. G. Street's father's farm, everything revolved around the sheep, and it is easy to understand his point of view. The annual cycle of sheep management was every bit as important as the arable one on many farms, and it made the shepherd the most important worker on a downland farm. Late winter and spring were the time for lambing; washing and shearing were completed between May and July, depending on type of sheep and locality, followed about six weeks later by dipping. Keeping the sheep fed was a constant activity: about half the arable on a

A sheep fold on downland.

farm on light land was growing feed for the sheep. A large proportion of these crops were then used for a characteristic feature of this style of farming, the summer folding of sheep. The sheep were fed on the crops of clover where they had been grown, part of the fine balance whereby sheep and cereals supported each other. Although it saved the work of cutting and carting the crop, it was an intensive procedure, involving the regular movement of hurdles around the field so that the sheep fed and manured each area in turn. In winter sheep were folded on the root crops, supplemented by purchased sheep feed.

Although pigs were a minority interest, prepared feeds were being made for them in the Edwardian years.

The pig, surprisingly perhaps, was little valued at this time, except on small farms, where it was often an integral part of farm management. For the farmer of a few acres, pigs could be a better choice than managing a few cows or sheep. Pigs were often kept alongside dairy farming; again the smaller the farm, the more likely this would be, especially if the farmer made cheese. But larger farmers took little interest in the animals: when they kept them it was only as a small sideline.

The farmyard pig – sow and litter.

41

BIBBY'S CAKELETTES
FOR SHEEP & LAMBS

THE BIBBY CAKE IN CAKELETTES
J. BIBBY

THE SHEPHERD'S WINDFALL

PASTORAL FARMING

THE NORTH and west were the pastoral zones of the country, where more of the land was in permanent pasture on which farmers bred or fattened cattle and sheep, and increasingly kept dairy cows. There was a great range in the nature of pastoral farms. In upland counties they could vary in size from a hundred acres or so to several thousand acres of hill grazing. Such extensive hill farms were on pastures that did not lend themselves to much improvement. Conditions for both stock and farmer could be exacting. This was the home of the tough hill breeds: the Herdwick sheep of the Lake District, the Blackface and the Cheviot. Shepherds were important on upland farms, and numerous: there could be two or three living in at the farm, and others accommodated elsewhere. Lower down, cattle could be kept, again often of hill breeds, such as the Galloway and North Devon.

There was greater prospect of improving the lower pastures with manure, artificial fertilisers and new seed mixtures. Where possible the pasture farm had some arable, even in the upland zone of England. Akeld Farm, in Northumberland, was very large, extending to 1,800 acres, of which 1,100 acres were hill grazing and the rest rich tillage land in the valley. The more lowland the farm, the more likely it was to have a portion of arable land on which to grow crops mainly to feed the livestock – oats, roots, hay – though some, such as potatoes, might be a cash crop as well. The stock these farmers kept were mainly to fatten or for dairying.

There was pastoral farming in the southern lowland zone of England as well, with a long-established tradition of specialised pastoral farming on the renowned rich grazing lands of Leicestershire and Northamptonshire. Farmers in this area had little arable, and had reduced its extent since the depression years. They bought in stock to fatten, much of it from far away, such as Wales and Ireland.

When Sir Daniel Hall visited the Vale of Pewsey he found an area that had become almost entirely pastoral since the 1870s. It was not the only example: there were many others where farmers had found greater

Opposite:
J. Bibby & Sons Ltd. were producers of prepared livestock feeds. This is an advertisement for their 'cakelettes' in 1912.

Cheviot sheep, one of the main hill breeds, at Akeld Farm, Northumberland.

economy by reducing arable to a minimum and concentrating on pastoral farming. Extending the grassland had not always been very systematic. Some land had been allowed to revert naturally to grass rather than being sown. But the Edwardian farmer was paying more attention to the improvement of pastures, with special seed mixtures available for different

Herdwick sheep at Rydal Farm, Westmorland, 1907.

The Galloway was a hardy Scottish breed of cow, kept by some farmers in the north of England, and among the types taken south for fattening on southern pastures.

situations, and fertilisers to increase yields. One of the farms in the Vale of Pewsey was S. W. Farmer, who had a total of 20,000 acres in 1914. One thing he shared with almost all of his neighbours operating on a much smaller scale was a concentration on dairy farming.

A byrewoman feeding the fat cattle on Akeld Farm, Glendale, Northumberland, featured in *Country Life* in 1901.

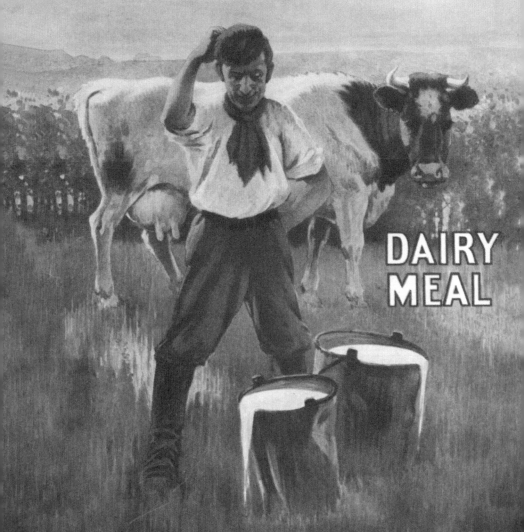

DAIRY FARMING

JOHN SHELDON was a professor at the Royal Agricultural College at Cirencester who wrote extensively on dairying. In 1908 he wrote of 'the exalted position which dairy farming fills today as compared with forty years ago'. The growth of dairy farming since the 1870s had been quite spectacular, one of the most successful ways in which farmers responded to the downturn in cereal prices. By 1911–15, there were 30 per cent more dairy cows on English farms. Not only that, but the nature of dairy farming itself had changed. Whereas butter and cheese had once been the principal produce, 70 per cent of the production from Edwardian dairy farms was sold as liquid milk. Some of that went to creameries for processing – there were new creameries established at Swindon, Aylesbury and Stafford, for example – but most was sold for consumption as milk.

Growing demand from the big towns and cities since the middle of the nineteenth century had made the milk trade a big business. By 1900, 53 million gallons of milk were arriving in London by rail each year compared with 9 million in 1870. London was by far the biggest market, drawing supplies from as far as 200 miles away, from farms in Cheshire, parts of Wales and Devon. Most of the capital's supplies came from closer to hand, from Berkshire to Somerset in the west, and from Staffordshire and Derbyshire in the Midlands. While London was dominant other major towns and cities also provided healthy markets – Manchester and Liverpool for the dairy farmers of Lancashire and Cheshire, the West Riding towns for farmers in the Yorkshire Dales. Farmers near to an urban market were sometimes able to set up their own retail operation, but most sold their milk to distribution companies.

The six counties of Cumberland, Westmorland, Lancashire, Cheshire, Derbyshire and Staffordshire were the major English dairying counties, accounting for more than a quarter of milk production in 1908, while Devon and Somerset were also important. All of these counties had an established pastoral tradition: dairy cows fed on the grass pastures, supplemented by fodder grown on the farm and purchased feed, were a feature of local

Opposite:
Bibby's dairy cattle feed advertisement.

Milking continued to be conduced mainly by hand during the Edwardian years. It was not necessarily performed out in the field, but to the photographer, this type of scene was more attractive.

farming. Pure pastoral dairy farming was in the minority, however. Dairying worked best in mixed farming, J. P. Sheldon thought. 'A portion of land under arable cultivation is very useful, and even indispensable, on a dairy farm,' he wrote. Joseph Stevenson at Baulking was a mixed farmer, in common with most dairy farmers: according to the official returns, 75 per cent of dairy cows in England and Wales were kept on mixed farms. Even in the most pastoral of situations a supply of home-grown straw, oats, roots, clover and

This scene of milking in the farmyard was reproduced as a postcard in 1907.

A farm near Penshurst, Kent, depicted in one of the many postcards published in the Edwardian years. A little donkey cart has some milk to be delivered.

other fodder crops was invaluable as feed and winter bedding. The mixed farm in any case made for a more varied occupation for the farmer, and gave him some additional possibilities for income.

Sheldon was not in favour of dairy farming on a purely arable farm, and only 2 per cent of dairy cows were kept on arable farms according to the production statistics. Arable-based dairying did have its advocates, however, and was a strong trend during the Edwardian years. It took dairy farming

Milking in the cowshed on a farm near Alton, Hampshire, 1910.

beyond its 'homeland', from the 'cheese' (the lowland vales) to the 'chalk' area of Wiltshire, for example, where land had been put down to grass to support the dairy herd. There were some unexpected places where dairy farming took off. Essex was the most prominent. Here was a mainly arable county where in the late-nineteenth century dairy farming became a major activity, supplying milk to London. Part of this was down to an influx of farmers from the West Country and Scotland who brought with them knowledge and experience of dairy farming – and often their complete farm of dairy cows – which they adapted to the dry climate and heavy clay soils of Essex. One of these farmers, Primrose McConnell, became well known because he wrote about his experiences, publishing *A Diary of a Working Farmer* in 1906. He brought his Ayrshire cows to an Essex farm, but not the practice of dairy farming, for his predecessor as tenant had also kept cows. On the estate of Lord Rayleigh dairy farming was taken up in a serious way. The estate's own farms were expanded as a dairy business, with a retail trade built up alongside the farming.

There was pasture on the dairy farms in these arable districts, such as Essex and light-soiled parts of Wiltshire, but the feeding of the stock was based mainly upon the arable and feed bought in. Primrose McConnell described a system on the Essex clays in which the crops were grown with as much intensity of production as they always had. Wheat still had an important a place in his farming, as much as on any farm in Essex. But there were important differences between this Edwardian farming and its Victorian predecessor. McConnell sold most of his wheat crop, but he did not depend upon it for his livelihood. Wheat now had a role supporting that of the dairy

Ayrshire cow, 'Dairymaid', prize-winner at the Royal Show in 1906.

A Shorthorn heifer, 'Sherborne Fairy', winner at the Royal Show in 1910.

cows, and he had the option of giving it to them as feed, along with the crops of beans, peas, clover and turnips.

Dairy farming was labour-intensive, with a steady daily round of feeding, cleaning and milking. It suited the small family farms, those that could be managed on the labour of the family only, or with a very small additional input. Official statistics confirm this close relationship between dairying and small farming. In 1908 the greatest density of stocking of dairy cows was on farms of less than 50 acres, with a high proportion kept on farms smaller than 5 acres. Dairy farming needed good buildings as well – cowsheds and milking parlours – but as long as the landlord was prepared to provide them, the business remained within the scope of a small farm. Of course it was perfectly possible to manage dairy cows on large farms, and at the opposite end of the scale from the farmer with a few cows on 5 acres was Frank Stratton in the Vale of Pewsey, who had a herd of 2,000 cows, while Lord Rayleigh's farms had nearly 6,000 cows by 1913.

Primrose McConnell kept the Ayrshire cows that he had grown used to in Scotland. The Ayrshire was one of the few pure dairy breeds in Britain. The predominant breed of cow, however, was the Shorthorn. More than 64 per cent of all the cattle in Great Britain in 1908 were of the Shorthorn type. Even though the emphasis of breeding for several decades had been on its beef qualities, it was also the principal breed for dairy farming. Another of the Scottish farmers in southern England, Mr MacArthur at Bottisham, Cambridgeshire, kept some Friesian cows alongside his Ayrshires. So did Edward Strutt on Lord Rayleigh's farms, but the Friesian or Holstein, later to dominate dairy farming, was still a novelty. A sign of things to come was the foundation of the British Holstein Cattle Society in 1911.

HORSE POWER, MANPOWER AND MACHINES

R. E. PROTHERO was a leading agricultural commentator who became President of the Board of Agriculture during the First World War. In 1901 he wrote in the *Journal of the Royal Agricultural Society of England*: 'Without the aid of mechanical innovation farming today would be at an absolute standstill.' The Edwardian farm's implement shed was well stocked, and the period had its own contributions to make to the development of mechanical innovation.

The horse was still the prime means of motive power on the Edwardian farm, and the head horseman (called carter, waggoner or ploughman in different parts of the country) was among the most senior employees. The farmer had to keep enough horses to meet the peaks of demand, principally the autumn ploughing. As a rule of thumb, he needed four horses to the hundred acres. They were managed as teams of two or four. The bigger farms and estates liked to keep quality horses of the recognised breeds of heavy carthorse – the Shire, Clydesdale and Suffolk. These were far from universal, however, and the farms of England and Wales were worked by quite a mixed bunch of mongrel horses, some heavy, others less so. Most of the implements could be handled with one or two horses, but on heavy soils three might be needed for the plough and the seed drill. The binder for cutting the corn was a heavy machine that almost always required three, and often four horses.

There were other sources of power. Steam had been introduced to the farm from the 1830s–40s onwards, and in the Edwardian years contributed as much as 20 per cent of the total power used. It was a sign of agriculture in recovery that contractors and farmers were buying steam engines again after a long period of dearth. Steam power was concentrated mainly on the work of threshing, and associated work, such as baling straw and hay. The steam plough was used on some of the large arable farms of the eastern and southern counties of England. After neglecting it during the depression of the late-nineteenth century, more farmers were beginning to turn to it again in the new century.

The Edwardian farmer had a new source of power available to him – the internal combustion engine. Stationary engines using coal gas had been

Opposite:
The oil engine was used to drive some field implements, such as the elevator, powered in this illustration by a Blackstone engine. Hampshire, c. 1908.

Corn-grinding mill, from a catalogue of Bamfords' of Uttoxeter.

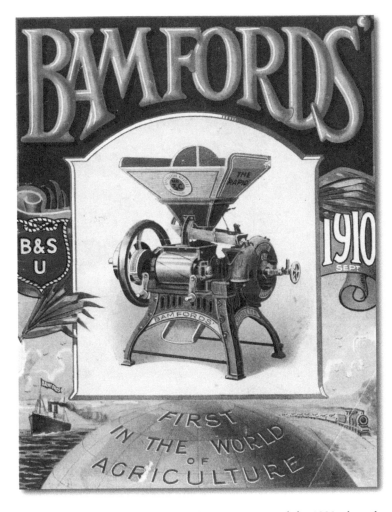

introduced as long ago as the 1870s, but it was not until the 1890s that oil and petrol engines were introduced, and these were the ones used most on the Edwardian farm. By 1910 there were about 16,000 of them at work on farms in England and Wales, still a relatively modest total compared with the number of farms, but it is reckoned that they were providing about 10 per cent of the power consumed in agriculture.

These engines were used mainly to drive the range of implements – chaff-cutters, cake-breakers, and grinding mills – that were collectively known as 'barn machinery'. They were for preparing animal feed. All were of a basic design, but the types available to the Edwardian farmer were updated, with greater capacity, improved rollers, cutters and safety guards.

What the farmer really wanted was a 'general-purpose tractor and engine for farm purposes', as the judges at a trial of engines organised by the Royal Agricultural Society of England put it. Those trials were held at Baldock, Hertfordshire, in 1910, and three steam tractors competed against two oil-

Steam cultivating in Suffolk. This set was about thirty years old when it was photographed in c. 1907.

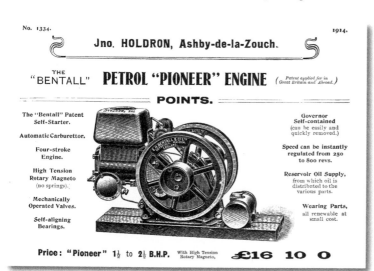

No. 1334.   1914.

Jno. HOLDRON, Ashby-de-la-Zouch.

"THE "BENTALL" PETROL "PIONEER" ENGINE (*Patent applied for in Great Britain and Abroad.*)

POINTS.

The "Bentall" Patent Self-Starter.

Automatic Carburettor.

Four-stroke Engine.

High Tension Rotary Magneto (no springs).

Mechanically Operated Valves.

Self-aligning Bearings.

Governor Self-contained (can be easily and quickly removed.)

Speed can be instantly regulated from 250 to 800 revs.

Reservoir Oil Supply, from which oil is distributed to the various parts.

Wearing Parts, all renewable at small cost.

Price: "Pioneer" 1½ to 2½ B.H.P. With High Tension Rotary Magneto, £16 10 0

E. H. Bentall & Co., of Maldon, was one of the leading producers of internal combustion engines for farmers during the Edwardian years. Their Pioneer engine was introduced at the end of this period.

The Ivel tractor, introduced in 1902, was of a novel lightweight design (the old design seemed like an oil-engined version of the traction engine). It had an engine of 8 hp, increased later to 14 hp. This is the 1903 model shown with its engine cover off drawing a binder. Its purchase price in 1903 was £300.

engined tractors built by Ivel and Saunderson. The judges recognised that the oil engine would 'best suit the requirements' of the farmer, but the tractors then available were some way from meeting them. Although the appearance of the Ivel in 1902 had demonstrated the practicability of a light, manoeuvrable vehicle, it and the Saunderson of 1904 were still more a sign of things still to come rather than their arrival. Both had reasonable commercial success, but sales in the hundreds rather than the thousands – not enough to transform the farming scene.

The process of improving farm implements had been going on throughout the nineteenth century, and the Edwardian period added its own contribution. It carried forward the latest stage in the mechanisation of the corn harvest with the adoption of the binder, which had been introduced to Britain during the 1890s and became dominant during the following decade. This was the machine that could cut the corn and tie it into sheaves, using a mechanism not unlike the sewing machine. Its predecessor, the reaper, only did the cutting, leaving workers to follow gathering and tying the sheaves. The binder, wrote Fream, 'is the result of the application of a larger number of mechanical principles than are to be found in almost any other machine used upon the farm.' The binder was the new machine for the Edwardian age, saving labour and allowing inroads to be made into the 20 per cent of the cereal crop that was still cut by hand in 1900. It also introduced a feature that became characteristic of the countryside – binder twine for tying sacks, which was also used as gate hooks and for all manner of purposes.

There were new machines to aid the hay harvest, such as the side-delivery rake, which was introduced in the early 1900s. It combined the two jobs of

Side-delivery rake made by Martin's Cultivator Co. of Stamford, 1909.

haymaking – that is, the turning of the cut hay to aerate it and dry it, and the gathering of the crop.

Primrose McConnell in his *Diary of a Working Farmer* in 1906, wrote of trying a new plough, the 'latest thing'. McConnell was evidently quite a one for new ploughs. This was the thirteenth of his career, he said, and he was not unusual in that. But this latest thing was another sign of a new age, for it was a chilled steel digging plough, of a type designed in America and introduced to Britain during the late 1880s. British manufacturers successfully made

The Albion binder was made in England, by Harrison McGregor of Leigh, and was recommended for the quality of its knotting mechanism. This illustration is from the firm's catalogue for 1910.

"ALBION" Nº 3 STEEL BINDER

SOLD BY

The swath turner with rotating cones to turn the hay was designed by Jarmain's of Haseley, Oxfordshire. This is the model sold in 1908.

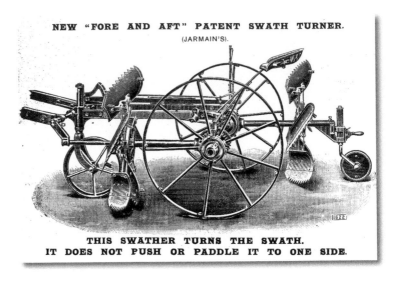

NEW "FORE AND AFT" PATENT SWATH TURNER.

(JARMAIN'S).

**THIS SWATHER TURNS THE SWATH.
IT DOES NOT PUSH OR PADDLE IT TO ONE SIDE.**

their own versions of the plough, but their complete dominance of their home market had been dented. The binder was another American machine, and American firms dominated the sales. Hornsby and Harrison McGregor, the two British firms making binders in the Edwardian years, were successful with their machines, but they required licence agreements or measures to circumvent patents.

Joseph Stevenson had a Massey-Harris cultivator, probably a new one, for he was proud enough of it to mention it by name in his day book. This was another North American product: Massey-Harris, a Canadian company, had not been trading in Britain for all that long. Massey-Harris offered most types of implement, but were best known for their harvesting machines, and for

The American type of plough favoured by Primrose McConnell. This is the original model made by the Oliver Chilled Plow Works.

BURMAN SHEEP SHEARS AT WORK.

It was easier to take a photograph in the open showing the shearing of sheep by hand-cranked machine clippers, but by 1907, when this was taken, best practice was to equip a shed with drive shafts from an engine.

Peter's Farm, Rusper, near Horsham, Sussex, from a postcard sent in 1910. The house, probably dating from the seventeenth century, is accompanied by a range of timber buildings, including a granary raised on staddles.

another new machine of the decade, the manure distributor. Shaped like a traditional farm wagon, this machine had belts and revolving tines which shot the manure out on to the field. Primrose McConnell was keen on yet another American innovation, the disc-coulter seed drill: it ran more smoothly than the 'antediluvian' Suffolk drill, he and other supporters argued.

Peters Farm, Rusper.

Among important smaller machines of the decade were powered clippers for shearing sheep, originating in Australia in the 1890s. At first they were operated by hand crank, which limited their appeal. Power from a gas engine improved their popularity, and they were becoming common during the 1900s.

Although dairy farming was growing in importance, it was almost untouched by mechanisation. Milking by machine was so uncommon that the text books barely gave it a mention. The vacuum milking machine had been invented as far back as the 1870s in America. More serious attempts to introduce it into British farming were made from the 1890s onwards, but with little success. Primrose McConnell was a great enthusiast for mechanisation, and spent £240 on a milking machine in 1903, but the results were disappointing. Despite making modifications to the machine, he found his milk yields fell. With such limited reliability, the majority of small farmers involved in dairying fought shy of using the milking machine until after the First World War. One new tool that did become established in the dairy was the cream separator, often powered by a small gas or oil engine. Specialist firms of suppliers to the dairy trade produced various measuring and testing tools, such as the hydrometer (which measured specific gravity) and butyrometer (which measured butterfat content) – as well as high-quality, hygienic dairy utensils, made of brass, tinned steel and galvanised iron.

A milking machine made by R. A. Lister & Co. Ltd. of Gloucester, c. 1914.

FARMING FIT FOR WAR?

THE OUTBREAK of war in August 1914 had an immediate effect on agricultural prices, especially cereals. The average price of wheat, 34s 2d a quarter at the beginning of August, reached 42s 2d in December and continued to rise into the new year. Prices of barley and oats also appreciated quickly. This was despite the good harvest that had been gathered in 1914: the simple prospect of diminished imports, which accounted for about four-fifths of British bread grain consumption, was enough to have great impact.

This was good news for the arable farmers of eastern England, who could sow wheat again, which they did. However, statistics for the acreage of crops show a distinct caution: sowing of cereals increased, but by less than might be expected from the rise in prices. Not until the government introduced its ploughing-up policy in 1917 was there a substantial change. It is easy to see why. Farming had established a pattern during the Edwardian years that was working; it had become remarkably productive and enabled farmers to reap some reward from the market as it stood. In 1914–15 there was insufficient incentive for farmers to disrupt that pattern. They needed some assurance that going back to cereals on a large scale was going to be worth it, and they received that from the Government in 1917.

Besides, in 1914 everybody expected the war to last no more than a few months. Farmers shared that view, and their newspaper, *Farmer and Stockbreeder*, was quite sure in November 1914 that agriculture would be little affected by the war. Farm labourers also shared that optimism, and left the farm for the army, often with great enthusiasm, in the expectation that they would have a short spell in France and be back home in a few months. Those in the reserves were called up, including the thousand men in the Waggoners' Reserve on Sir Mark Sykes's estate at Sledmere in Yorkshire. That immediately left a large hole in the local labour force, but everywhere farms soon felt the effect of their workers' absence. Along with the reserves went the volunteers, responding to Lord Kitchener's appeals. By July 1915 an estimated 150,000 men had left farming for the forces, about 15 per cent of the pre-war total. Making up for that loss was to be one of the main problems for wartime farming.

Edwardian farming could hardly be said to have been equipped for war. In 1914 the majority of the nation's food came from overseas, while English farming's emphasis on livestock products did not meet the immediate needs of wartime, for which a ready supply of bread grains was wanted. There were positives, however, for farming had restored its productivity during the Edwardian years and was in a reasonably sound condition. The output of agriculture in England and Wales in 1909–13 was little different from that of 1867–71. The nature of production had changed, but the total was the same. The greater question was going to be whether, having disturbed the equilibrium so painfully gained before the war, farming would be able to re-establish itself afterwards.

A ploughing match in full swing at Cippenham Court, near Slough, in 1909.

FURTHER READING

Armstrong, Alan. *Farmworkers: a social and economic history 1770–1980.*
Batsford, 1988.

Collins, E. J. T. (ed.) *The Agrarian History of England and Wales. Volume VII,
1850–1914.* Cambridge University Press, 2000. (An extensive account,
in two volumes.)

Fream, W. *Elements of Agriculture.* 7th ed., John Murray, 1905. (The text
book for the Edwardian farmer.)

Hall, A. D. *A Pilgrimage of English Farming 1910–12.* John Murray, 1914.

Horn, Pamela. *The Changing Countryside in Victorian and Edwardian England.*
Athlone Press Ltd., 1984. (An introduction to the subject.)

Hudson, W. H. *A Shepherd's Life.* Methuen and Co., 1910. (Classic account
of work on a downland farm.)

McConnell, Primrose. *Diary of a Working Farmer.* Cable Printing and
Publishing Co., 1906.

Moffitt, John. *The Ivel Story.* Japonica
Press, 2003.

Orr, J. *Agriculture in Berkshire.* The
Clarendon Press, 1918.

Rider Haggard, H. *Rural England.*
2 volumes, Longmans, 1902.

Street, A. G. *Farmer's Glory.* Faber and
Faber, 1932. (The first section
describes his youthful years on his
father's Edwardian farm.)

Wright, R. P. (ed.) *The Standard
Cyclopaedia of Modern Agriculture and
Rural Economy.* London, c. 1905–14.

A Hampshire
shepherd moving
hurdles around the
fold. A photograph
from c. 1906.

INDEX

Printed and bound by CPI Group (UK) Ltd, Croydon, CR0 4YY

11/10/2024

01043560-0015